Contents

In Meadows and Fields 2
Butterflies 4
The Mole 6
The Rook 8
The Crested Lapwing 10
The Old World Harvest Mouse 12
The Magpie 14
The Red-backed Shrike 16
The Common Shrew 18
The Meadowlark 20
The Weasel 22
The Little Owl 24
The Ladybug 26
The Colorado Beetle 28
The Starling 30
The June Bug 32
The Brown Hare 34
The Stonechat 36
Spiders 38
Snails and Slugs 40
The Earthworm 42
The Toad 44
The Blindworm 46
The Countryside through the Seasons 48

Library of Congress Cataloging-in-Publication Data

Cuisin, Michel.
  [Vie secrete des betes--a la campagne. English]
  Animals of the countryside / Michel Cuisin ; English translation by Jane Clothilde Lamb ; illustrations by Carl Brenders.-- U.S. ed. / edited and adapted by Joanne Fink.
    p. cm. -- (Nature's hidden world)
  Translation of : La vie secrete des betes--a la campagne.
  ISBN 0-382-09465-4 : $13.96. ISBN 0-382-09467-0 (pbk.) : $5.75
  1. Animals--Juvenile literature. I.Brenders, Carl. II. Title. III. Series.
QL49.C8513 1987
591.5'264--dc19

87-18627
CIP
AC

© 1982 Librairie Hachette, Paris. Originally published in French under the title *La vie secrète des bêtes: A la Campagne.*

© 1987 English text Silver Burdett Press.

All Rights Reserved.

First published in the United States in 1987 by Silver Burdett Press, Morristown, N.J.

U.S. edition edited and adapted by Joanne Fink

Nature's Hidden World
Animals of the Countryside

Michel Cuisin

English translation by Jane Clothilde Lamb

Illustrations by Carl Brenders

SILVER BURDETT PRESS

In Meadows and Fields

The abundance of animal life in the open countryside depends on the conditions that exist there. A concentration of plants of the same species such as corn, wheat, potatoes, and beetroot over hundreds of acres of land encourages an increase in the population of certain animals, particularly insects. They are able to reproduce in great numbers because of the huge quantity of food that farmers make available to them—usually without meaning to do so. Thus, on one hand, farming has increased the number of insects while on the other, many species of insects have begun to disappear due to the use of insecticides and the year-round work done in the fields. So, the diversity of animal life has actually decreased.

In a European forest there may be as many as thirty-five to forty varieties of nesting birds, while in nearby fields there are usually only about six including the gray partridge, quail, corn bunting, kestrel falcon, tree sparrow, and woodpigeon. Several of these species nest in trees or buildings. The most commonly found mammals are field mice, small rodents, hares, and small game animals.

Some creatures are more noticeable than others, simply because of the damage they cause. However, they are not the only creatures found in the countryside. There are many lesser known ones that play important and helpful roles. Some are prey for predators, while others transfer pollen from one flower to another. Others help make the soil fertile or are parasites on other species that are harmful to plants. Then, there are the millions of tiny worms in the earth helping to make the soil more fertile.

An Enormous Network
Hedges (boundaries or dividers formed by rows of shrubs or low trees that have been planted close together) do not occur naturally. They are found mainly in Europe and have been planted deliberately. Research has shown that hedges

are important, especially near the coast. They help to reduce the speed and drying effect of the wind, keep temperatures warmer, and guard against floods. Like forests, hedges stop water from running off too quickly and causing soil erosion damage.

Many animals live and feed in hedges and the surrounding meadows and fields. With the thousands of animal and vegetable species that live in the thick hedges, they could be called miniature wildlife preserves. Hedges provide animals with protection from the sun, rain, and the wind. Some forest animals, like the blackbird, song thrush, robin, field mouse, and wild rabbit, also make their homes in hedges.

Recently, many hedges have been destroyed to increase arable land and to make space so that farm machinery can be used economically This of course has upset the lives of many animals and has even caused some to disappear.

Butterflies

The peacock butterfly is found throughout Europe and eastward to Japan wherever nettle bushes grow—in gardens and meadows, near hedges and lakes. Some people even plant nettles in their gardens to attract this beautiful butterfly. Peacocks are found near nettle bushes because the female butterflies lay their eggs on these bushes. Nettles are the favorite food of the caterpillars when they hatch from the eggs.

The top sides of the peacock butterfly's wings are brightly colored with round spots on them called "eyespots." These eyespots look very much like the spots on a peacock bird's feathers. While in flight these brightly colored wings make it very easy for a bird to spot this butterfly and attack it. However, while at rest on a plant or twig, the butterfly uses this sometime hazard to its advantage. The undersides of the peacock's wings are dark. It holds its wings up, so only the dark sides are seen and sits there looking more like a leaf than a butterfly. If threatened by a bird the butterfly will spread its wings quite suddenly. By opening its wings very quickly and exposing the top side, the peacock butterfly scares off a potential enemy by startling it with a burst of color.

In the spring and late winter red admiral butterflies (bottom) are seen in the countryside when they emerge from their winter hiding places. In the spring the females fly to northern areas to lay their eggs. In the fall the butterflies that have hatched from these eggs fly to warmer climates in the south. Red admirals can be found in meadows and grassy places as late as October or November. They are common in many parts of the world including North America and Europe.

Butterflies—some kinds are great travelers—are sometimes tagged by researchers so their movements and contacts with other butterflies can be studied. Tiny labels with a number and the abbreviated address of the research institute studying them are glued onto the front part of each butterfly's wings. Eighteen hundred little tortoiseshell butterflies were tagged and released in early summer. Ninety-two of them were found later, one as many as 251 days later. Some were found almost 100 miles from where they had been released originally.

Wingspan: little tortoiseshell, 1.7 inches; red admiral, 2 inches; peacock, 2.4 inches.

▲ *The metamorphosis of the little tortoiseshell. The little tortoiseshell reproduces two or three times a year. Eggs are laid in June/July, August/September, and October/November. From left to right, the various stages of a butterfly's life are:*

▼ *Top left: an enlargement of a butterfly head showing one eye, the base of the antennae and probocis. Top right: the top of the head and compound eyes. Bottom right: detail of a wing showing microscopic scales. Bottom left: a hairy leg.*

egg, caterpillar, chrysalis, and butterfly. As many as 30,000 caterpillars to 27,000 square feet can be found in places overgrown with nettles!

▼ *One of the meadowland butterflies in France is the burnet. It has bluish black, red-spotted wings and a black body. Burnets feed on the flowers of crown vetch, clover, and bird's foot trefoil and draw little attention because they are not as graceful as other species.*

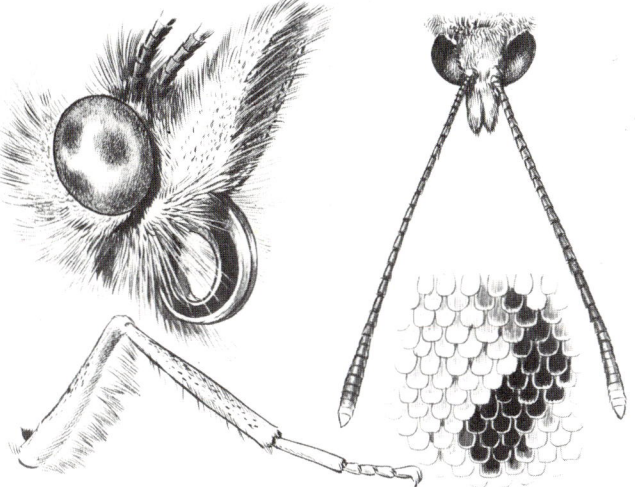

▼ *The little tortoiseshell lays pretty, ribbed eggs on nettle shoots. As many as two hundred are crowded together. After eight to ten days, young caterpillars emerge. At first, they stay together and eat up all the nettle leaves before turning into chrysales.*

The Mole

A little earth-mover

The female mole uses its shovel-like forepaws to dig a network of underground tunnels from forty-three to fifty-five yards long. Tunnels dug by males range in length from fifty-five to one hundred sixty-five feet in length. Moles also use their paws to push excavated dirt up out of their tunnels to the surface of the ground. That is how molehills—little mounds of earth—are formed, and, this is when the mole is seen above ground. The mole also surfaces in dry weather during the summer or winter; at both times apparently looking for prey.

The mole is almost blind and feels its way around in the dark underground by using its highly developed senses of smell and touch. It also has an astounding capacity to sense vibrations and can escape if it feels the earth being disturbed. It moves along with its tail touching the roof of its tunnel and its body snugly filling up the center, always in contact with the earth around it.

Greedy for worms

Before winter sets in, the mole stores up a large supply of earthworms. As many as 470 have been found in one tunnel! The mole paralyzes the worms to keep them from escaping by biting the front of their bodies. It also stocks up on millipedes, slugs, and insect eggs—its basic diet along with earthworms. The mole usually catches maybeetle larvae, click beetle eggs, and large mosquitoes. One mole can eat up to one and three-quarter ounces of food daily. It cuts through roots, but does not eat them, when digging tunnels. Moles are considered a nuisance in gardens and fields because they destroy seedlings and roots.

Length: 5 to 5½ inches plus a 1 inch tail.
Weight: 2.8 to 4.2 ounces.
Lifespan in the wild: about 3 years.

▲ A mole digs its tunnels by using one forepaw and then the other. Contrary to popular belief its snout, which is very sensitive, is not used for digging. A mole can dig through about thirteen pounds of earth in twenty minutes. This would be the same as a 165-pound coal miner moving twelve tons of coal in an hour—an impossible task for a human being. At the far right: a six- to twelve-inch molehill in a field.

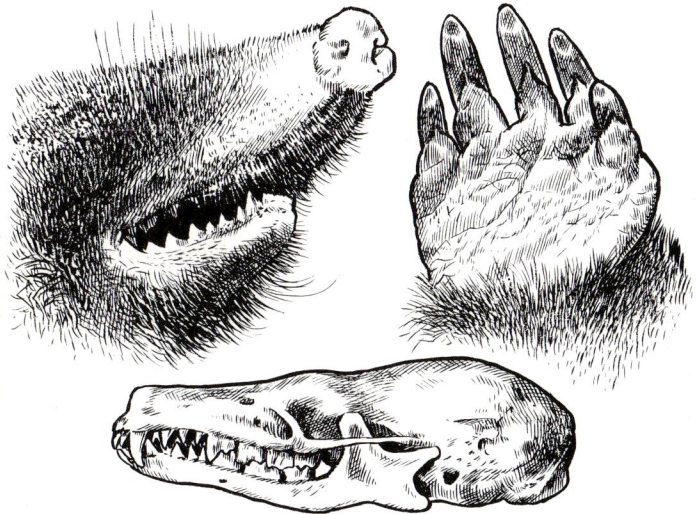

▲ The tip of the mole's snout has organs that are sensitive to smell and touch. Right: a forepaw viewed from above. It looks somewhat like a human hand because it is hairless. The mole's head is elongated like a shrew's head. There are forty-four teeth in a mole's jaw.

▲ The mole mates in March or April, and three or four baby moles are born in May or June. At birth they are hairless. They show a coat at fourteen days and open their eyes at three weeks. Sometimes moles make their nests in molehills that are larger than the usual size.

▼ This mole is devouring a June bug. Moles eat many June bugs, which are harmful to young plants. A hungry mole can eat a half an ounce of earthworms before eating its fill. In captivity, moles have been known to go twenty-four hours without eating. A mole can swallow a large worm weighing a quarter of an ounce in a few seconds.

The Rook

Peaceful invaders

Every year in autumn, hundreds of thousands of rooks invade France. They come from central and eastern Europe flying closely together over Switzerland and Germany; in fifteen minutes 5,000 have flown over Fribourg-en-Brisgau in southern Germany. In one afternoon 12,000 have flown over the Black Forest near the Vosges Mountains, and 10,000 to 12,000 have flown over the Swiss Jura Mountains in two days. These black birds fly at an average speed of almost 40 miles an hour.

Usually rooks travel slowly, taking about two and a half months to cover a little over 1,250 miles. These birds fly over the countryside a few hundred yards in the air and can be identified by the croaking sounds they make before coming down to land in trees for the night.

A rook dormitory

Rooks are gregarous birds. This means that they nest in flocks. They are members of the crow family. Like crows in North America, the rooks' only enemies are people because these birds eat up grain in fields. In some regions, they are hunted and shot. However, even if they do eat grain, they also eat many other insects that are harmful to planted fields–June bug larvae, weevils, locusts, and crickets. Their diet consists mainly of these creatures.

In winter, rooks look for food in plowed and stubbly fields, and in manure heaps. At nightfall, they gather into one flock and head for the woods where they are living for the season. These "dormitories" sometimes "house" as many as 10,000 rooks along with jackdaws, black crows, and other, similar birds.

Rooks mate in plains or low plateaus less than a few hundred feet in altitude. They are native to Europe and Asia.

Length: 17 to 18 inches.
Weight: 14 to 18 ounces.
Wingspan: approximately 35 inches.
Average lifespan in the wild: 10 to 11 years, maximum 20 years.

▲ *The rooks pair off into couples by doing a mating dance that involves a series of movements, attitudes, cries, and songs. Above, part of the rook's mating dance. Left: the female begs for food by spreading and flapping her wings and letting out a cry similar to that which the fledglings make after they hatch. Right: After her performance the male gives the female a ''present'' of an insect or a seed.*

▼ *Top: The fledgling has feathers at the base of its beak that fall off when it is adult. Left to right: A young rook, a six-month old rook starting to molt, and an adult. Bottom: The carrion crow (left) resembles the rook (right) but has different habits.*

▼ *Rooks nest in colonies. Hundreds, even thousands of rooks build their nests close together in parks, woods, or poplar trees. When there are no leaves on the trees, their nests can be seen from far off.*

▼ *Rooks build nests in large trees. Sometimes there are forty to sixty nests in the same area. In March, the female lays three to five greenish-gray eggs and sits on them for sixteen to eighteen days. The young learn to fly when they are a month old. Here, an adult is feeding its babies.*

The Crested Lapwing

Aerial stunts

After flying straight up into the air for about ten yards, the male lapwing then plunges back down to earth, flapping his wings very quickly and making a strange noise. Only the male can produce this noise, for his wings are an inch and a half longer than the female's. While the lapwing does his aerial stunts during mating season he lets out sad, high pitched cries to show he is excited and also let his neighbors know where he is.

Both the male and female lapwings have a rather slow, awkward way of flapping their wings while flying. This is where their name "lapwing" has come from. The latin name for these birds, *Vanellus vanellus*, comes from the latin word *vannus* meaning fan, and also refers to their wingbeat.

Crested lapwings are members of the plover family. They are a common sight in France during the winter months, migrating from northern and central Europe in flocks of hundreds or even thousands to enjoy the warmer weather. During migration lapwings generally fly at an altitude of less than 1,700 feet, but they can fly as high as 13,000 feet to avoid clouds. Their maximum speed is about forty miles an hour.

An original method

Lapwings hunt for food in damp fields and prairies. They prefer natural, treeless sites with low grass. They locate some of their prey by listening for noises. Others they find in a rather curious way. While standing on one leg, they shake the other back and forth very quickly to loosen surface mud or move loose earth. This allows them to peck at the insects hidden there. They like crickets, ants, weevils, beetles, and flies as well as spiders, worms, and mollusks.

Length: 12 to 13 inches.
Weight: 7 to 9 ounces.
Wingspan: 27 inches
Lifespan in the wild: 25 years maximum.

▲ Lapwings mate in March and April. In the top left picture a male and female are in the foreground. In the center and right pictures the male takes off and turns around just above the head of his mate. In the bottom row, the aerial stunt continues while in the background the other lapwings have not yet formed couples. Finally, the male lands near the female who has begun to make a nest with dry grass she has gathered in her beak.

▲ Top: The wing of a male lapwing, below it, the wing of a female. Males can make their strange noise due to a larger wingspan. The male's crest measures five inches and the female's measures three inches. Like other small wading birds the lapwing's feet have a small hind claw.

▼ The lapwing sometimes begins nesting when there is still snow on the ground. That is why its clutch is sometimes killed by the cold. The bird has little defense against the elements, but can protect its eggs and young from live enemies quite effectively. If the bird feels that its nest is being threatened she will leave the nest, limping and faking an injury to draw enemies away.

▲ The female lays four pear-shaped eggs with greenish-brown spots. The eggs are laid in April or May and hatch twenty-six to twenty-nine days later. An hour or two after hatching, the chicks begin tottering around and are actually able to leave the nest after only a few hours.

The Old World Harvest Mouse

The Old World harvest mouse is commonly found living in fields of tall grass in Asia and Europe. It has a long, prehensile tail (one that can be used for grasping) that it uses to cling tightly to stalks and stubble. Its hind paws have a "thumb claw" which help it to hold on. The harvest mouse is a good climber because of its small size and light weight. This very nimble little creature is generally bright yellow or reddish brown in color.

When frightened, the harvest mouse stays very still, hoping not to be seen, or it scampers off quickly. It has a very good sense of hearing, but poor vision. The Old World harvest mouse is active mainly during the daylight hours. As winter approaches, it stops scurrying around in the thick underbrush to seek refuge in burrows dug by other rodents like the field mouse.

A similar creature, the American harvest mouse can be found in this country. It differs somewhat from the Old World harvest mouse in that it is pale brown in color, spends most of its time in places where the grass is short, and is basically a nocturnal animal. The American harvest mouse can be as much as one inch larger than an Old World harvest mouse, but they are both among the smallest members of the rodent family.

The Old World harvest mouse stays close to home unless disturbed. In summer it will rove over an area of five to seven hundred square yards. The young, however, do leave their parents' territory. Soft-bodied insects, small fruit, and principally seeds are the staple diet of this tiny rodent. It can easily climb to the tip of a stalk to gnaw on the barley or wheat there. In some years there may be as many as two hundred harvest mice on two and a half acres of land. However, the conditions necessary to produce such large numbers of them only occur in certain areas and do not last long. This is the reason why farmers do not find these creatures to be a nuisance.

Length: 2 to 3 inches plus a 2 inch tail.
Weight: .2 ounces.
Lifespan in the wild: 18 months.
Lifespan in captivity: 5 years.

▲ The harvest mouse builds its nest in grain fields, reeds, grassy places, and meadows. Its nest is attached to a plant about eight to twelve inches above the ground. The female builds the nest just before giving birth in June. An architectural wonder, the nest is globe-shaped measuring three to four inches in diameter. It has one room and a side entrance. From left to right, the nest being built. It is woven out of vegetation.

▼ Several kinds of carnivorous animals are a threat to the harvest mouse. Here, a kestrel falcon is preparing to swoop down on two mice it has just spied. The harvest mouse's enemies include the barn owl, the tawny owl, and the weasel. Rain and cold also kill young and old alike.

▼ Below top left: a harvest mouse's skull. Bottom left: a field mouse's skull. Right: the harvest mouse's "hand" gripping a vertical stalk. The tip of the tail is rolled around a stalk like a clamp to keep the creature from falling when it moves around above the ground.

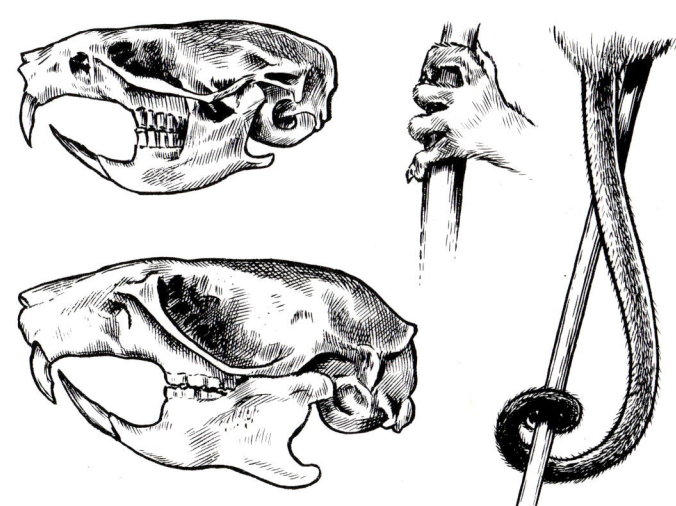

▼ A female with her young inside a nest. There are a maximum of three litters a year. At birth, the young weigh less than one gram and only open their eyes after eight or nine days. They become independent when they are two weeks old and are sometimes chased out of the nest by their mother. They are born between June and October.

The Magpie

Chattererers or robbers?

Legend has it that the magpie spends most of its time stealing precious objects that shine in the sun and hiding them. Supposedly, it takes advantage of open windows to get inside and steal things. This is how the magpie got a reputation for being a robber. In fact, only tame or captive magpies pay any attention to shiny objects. Only rarely is a magpie in the wild interested in glittering, metal objects like the one shown in the picture stealing a camper's watch. However, the magpie does deserve its reputation as a chatterer and can be heard all year round especially in late winter or early spring. Although the magpie is one of the "songbirds" of the sparrow family, the magpie, like its cousin the crow, does not have a beautiful voice and cannot sing. The magpie makes loud, low-pitched noises. The magpie chatters the most during mating season when the birds pair off into couples. Since there are usually about twenty involved in the mating dance their cries can be heard from quite a distance away.

A regular fortress

Spherical in shape, the magpie's nest is like a fortress made out of dry, thorny twigs. Built in poplars, or in large bushes, the nest has a bowl-shaped bottom made of earth covered roots. The nest has a roof made of twigs and entrances on the sides.

The magpie can be found on farmland where fields and meadows are bordered or divided by thickets, woods, or clumps of trees. The bird does not fear people and often goes into villages or towns if there are large trees suitable for nesting.

Length: 16 to 16½ inches.
Weight: 5½ to 9 ounces.
Wingspan: 19 to 21 inches.
Lifespan in captivity: 20 years.

▲ In autumn and winter, magpies gather in flocks of as many as one or two hundred birds to seek shelter for the night in groves or woods. In the morning, they leave the "dormitory" to look for food in the countryside (top left). These magpies have found a dead rabbit (center) but squabble over the booty. Magpies feed on insects, dead animals, eggs, young birds, worms, snails, seeds, and fruit.

▼ In March or April the birds form couples and start to build their nests. It takes from one to four weeks for a nest to be completed. Each nest serves for only one season. It is not rare, even in winter, to see magpies in pairs. Some people believe that the birds pair off and stay together for life.

▼ Top row below: two large feathers taken from a magpie's wing. The black part of the feathers is stronger than the white because pigment makes the feather barbs harder. Bottom row: the feathers at the base of the beak.

▼ This young magpie has left the nest but still opens its beak to be fed. It is about twenty-five days old. In April or May, the female lays six to eight eggs and sits on them for seventeen to eighteen days. When the fledglings leave the nest, their tails are shorter than adult's tails.

The Red-backed Shrike

The shrike assembles a motley collection of June bugs, crickets, small rodents, and young birds by impaling them on thorns or wedging them in the fork of a branch or a tree before eating them. For this reason shrikes are sometimes called "butcher-birds." Once its prey is impaled the shrike can then use its bill to pick the animal apart and eat it. More than three-quarters of the shrike's diet is made up of large insects. Male and female shrikes look quite different from each other (see male on top left). The females look more like fledglings that have just left the nest.

A great traveler, the red-backed shrike leaves Europe in late summer to begin its night-flight migration that takes it as far as the southern Sahara in Africa. According to calculations, it takes this bird three months to reach Africa by flying for a total of thirty nights and resting for seventy. The journey back, however, is quicker. It takes the bird only two months to reach a nesting place in Europe. In autumn, these birds fly through Greece, Libya, and Egypt. In spring, they follow a different route through Syria and Turkey to arrive in Europe in early May. Those nesting in northern Europe cover approximately 7,500 miles round trip. The shrike has become rarer in Europe during the last thirty years. This is apparently due to a lack of hedges, changing weather patterns, and the widespread use of chemical insecticides that have all but eliminated the large insects the birds feed upon.

In this country there are two species of shrike. The northern shrike is similar to the red-backed shrike and can be found from the northern United States into the most northern parts of Canada. The loggerhead shrike is found throughout the United States, north into the southern part of Canada, and south into Mexico.

Length: 7 inches.
Weight: 1 to 1.2 ounces.
Wingspan: 11 inches.
Lifespan in the wild: 6 years.
Lifespan in captivity: 8 years.

▲ A small field mouse has just been captured by a male shrike. The bird seizes its prey firmly (above left) in its hooked beak and carries it off to its "pantry" where the prey is impaled (center). Next to the field mouse, on other thorns, are earlier victims (above right): A frog and an insect waiting to be eaten. These food stocks (the bird can have three or four of them) are several yards away from the nest.

▼ Top left: a clear view of the hooked beak and the little "tooth" on the edge of the upper jaw. Below: head of male with black band. On the right: the tail of the male. When angry, the bird moves its tail up and down and from side to side.

▼ Although the shrike eats other animals, it is also the prey of certain predators, both birds and mammals. Its main enemy is the sparrowhawk. In the picture below, a weasel has grabbed a baby shrike. Sometimes a gray cuckoo will take over a shrike nest and destroy the eggs.

▼ The female lays five or six highly colored eggs. She sits on them for fourteen to sixteen days. The eggs can be yellowish, greenish, or reddish. The spots are at the thick end of the egg.

The Common Shrew

The common shrew is just a few inches long and weighs only ounces, but for its size it is one of the more vicious members of the animal kingdom. A very active creature, the shrew is on the move day and night looking for food. To keep up its rapid pace (it moves about five times faster than a rodent like the field mouse) the shrew eats an amount equal to or greater than its own body weight. The shrew has very good hearing, and some kinds locate their prey with a sense of echolocation like bats and dolphins. The shrew likes to hunt at night most of all. All the tiny creatures it comes across are its prey—earthworms, slugs, spiders, snails, and woodlice. The shrew will also feed on dead birds and animals.

A shrew cannot go for long without eating. After several hours without food some shrews will die. When food is hard to locate, or scarce, they will even feed on other shrews. The common shrew does not hibernate in winter. Its need to eat continuously forces it to face the cold.

Born in the spring, a shrew reaches adulthood in a year, reproduces between May and October, and dies shortly thereafter. They generally live alone and spend time with other shrews only for mating purposes. A shrew's lifespan does not exceeed two years because of the many hazards it must face—the cold, famine, and a variety of predators. At night, barn owls pounce on them as they move through fields, meadows, and clearings. Its other enemies include the weasel and the polecat. Some animals stay away from the shrew because it can secrete foul odors from special glands.

During the day, the common shrew finds shelter in abandoned tunnels made by field mice or moles. Sometimes it will dig a small shelter of its own under a pile of dead leaves.

Length: 3 inches, plus a 2-inch tail.
Weight: 2.3 to 5 ounces.

▲ Shrews are not rodents as one might assume from their appearance. Zoologists put shrews into the insectivore group of animals because they are insect-eating mammals. The cricket in the picture above has ventured from its hole. The shrew draws near to the cricket, breaks its shell, and swallows it in almost one gulp. Satisfied with its meal, the shrew departs, leaving behind only a few appendages as proof of the drama that has just occurred.

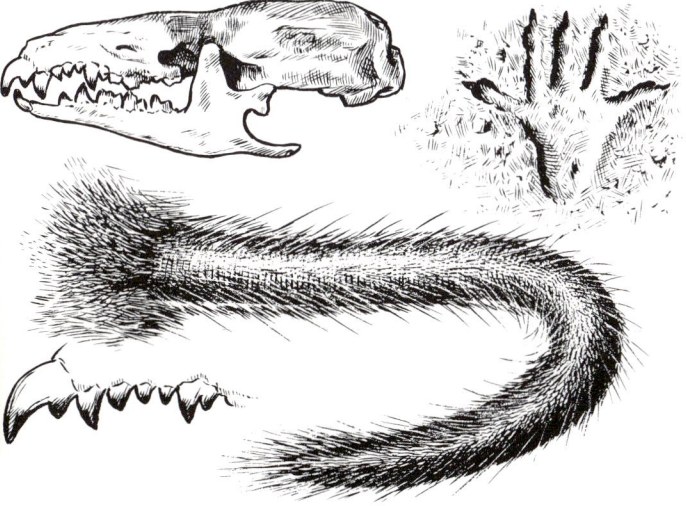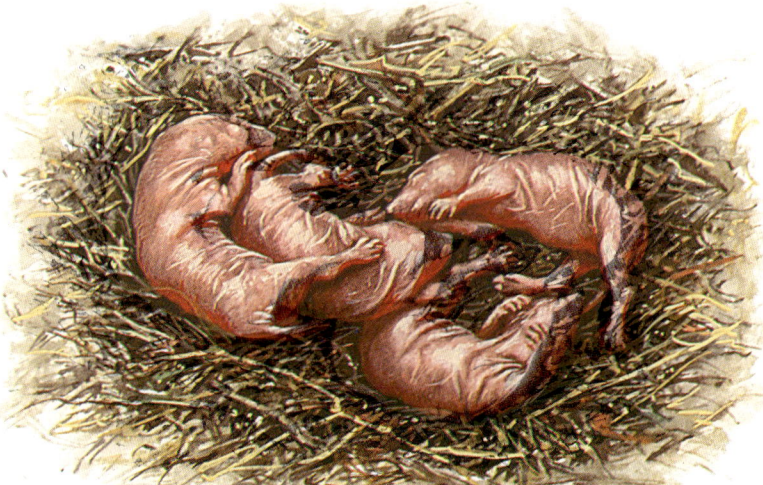

▲ The common shrew has thirty-two pointed teeth with red tips. It has a very flat skull. It rarely leaves tracks even on soft ground (top right) since it is such a lightweight animal. Its velvety tail is usually dark on the top and light on the underside.

▲ At birth, young shrews are hairless, blind, and weigh just half a gram. When they are eighteen days old they weigh about a quarter of an ounce and their eyes begin to open. They stop suckling at twenty-two days. Shrews nest under thick grass or fallen logs.

▼ The young of the white-toothed shrew, a cousin to the common shrew, can often be seen following their mother in a chain, one after the other. The first hangs onto the fur at the base of the mother's tail with its teeth, and the others follow suit behind. The scene resembles the procession of a slow moving caravan.

The Meadowlark

A tireless singer

The meadowlark starts to sing more than an hour before the sun rises. It will then sing for hours, stopping only occasionally as it flies over grain fields. From late January to mid-July, the meadowlark's continuous song can be heard. It becomes more intense in April and May as the bird settles in its mating territory. Flying straight up into the air, from 150 to 300 feet high, the meadowlark sometimes sings for a full twenty minutes without stopping. The song the birds sing while in flight, flapping their wings as many as twelve times a second, is different from the song they sing while at rest.

In the countryside meadowlarks can be found in marshes, plains, and mountain pastures up to an altitude of 8,000 feet. They prefer to live in alfalfa, clover, or grain fields, and only nest in places where the vegetation is not over eight inches high. They feed mainly on insects that we find annoying or harmful, and also on weed seeds.

Many enemies

The meadowlark has many enemies both on the ground and in the air. Its enemies include weasels, ermines, hedgehogs, foxes, and marauding cats. Carrion crows, magpies, and buzzards steal meadowlark eggs and fledglings right out of their nests. Falcons and sparrowhawks prey on the adult birds.

On the ground the meadowlark takes little tiny steps on its long-clawed feet. Its rear claw, ranging in length from half an inch to almost an inch, is longer than the three front claws.

In the United States there are two kinds of meadowlarks—the eastern meadowlark and the western meadowlark. These birds are quite similar to each other, but the western meadowlark is considered to have a more pleasant song.

Length: 6½ to 7 inches.
Weight: .8 to 1.7 ounces.
Wingspan: 12 to 14 inches.
Lifespan in the wild: a maximum of approximately 10 years.

▲ While the female builds its nest and sits on the eggs, the male sings. He rises in a spiral into the air and continues to sing as he flies back and forth over his territory. He zigzags back down to within a few yards of the ground and then, silently, drops like a stone into the grass and disappears.

▲ The meadowlark has a small crest on its head that is invisible when it is down (top left) but rises when the bird becomes excited (top right). Bottom left: the white strip behind the wing can be seen at a close distance when the bird takes off. Bottom right: the meadowlark's leg and foot.

▲ In April the meadowlark nests in hollows in the ground where the grass and weeds are thick. The female lines the nest with grass and small roots, and conceals it with a partial roof of grass and twigs. She lays three to five white eggs with rust speckles. These fledglings wait to be fed insects.

▼ Meadowlarks often take dust baths. They find a place where the earth is fine and sandy. There they dig a little hollow, wiggle around in it so the dirt works its way into their feathers, and then preen. This is probably the way the birds rid themselves of parasites.

The Weasel

When fighting an animal larger than itself a weasel is not necessarily the underdog for it is a notoriously fierce, but small carnivore. A weasel will pounce on its enemy, and try to administer a powerful bite at the nape of its neck, near the base of the spinal cord. The weasel will put up an incredible fight, and frequently come out the winner. It is not invincible, however, and other carnivorous (meat-eating) creatures do capture it—foxes or the great horned owl, its main enemy.

Weasels are not even frightened of humans, and when threatened will bite at their legs. Their numbers vary depending upon the abundance of prey in their territory. It has been estimated that in one year six weasels can kill an average of 2,000 field mice—rodents that can cause considerable damage to grain or alfalfa crops. In spite of this aid to the farmer, weasels are considered to be a nuisance. They are able to squeeze into the tiniest of holes and raid chicken coops, killing more chickens than they could ever eat.

The weasel is active both at night and during the day.

It moves in bounds and will sometimes stop and rise up on its hind legs to take a look around. In an hour it can cover just under a mile. It usually hunts over an area of thirty-five acres. The continuous activity of the weasel may shorten its lifespan, for on the average it lives for hardly two years. It is estimated that seventy-five percent of the weasels alive are born during the year, and only one out of eighty or ninety young weasels lives for more than two years.

There are about ten different kinds of weasels. Three are found in North America—the least weasel, the long-tailed weasel, and the short-tailed weasel. During the winter in cold climates the fur of many weasels changes from brown or reddish-brown to white, to camouflage it in the snow. When wearing its white winter coat the short-tailed weasel is called an ermine.

Long-tailed weasel length: 12 to 18 inches with a 5-inch tail.
Weight: 12 ounces maximum.

▲ This weasel, on its hind legs, has just found a field mouse's burrow. The weasel's body and short legs enable it to squeeze into the tunnel and hunt down the field mouse. The weasel follows the mouse through the tunnels, seizes it, and then drags its catch outside to eat it. Field mice account for one-half to three-quarters of the weasel's diet. It also eats earthworms, shrews, rabbits, small birds, frogs, lizards, eggs, and snakes.

▼ This rabbit is trying to escape from a weasel, but the weasel jumps up and grabs it by the neck. Weasels are excellent hunters, making use of their keen sense of smell and good eyesight.

▼ Weasels are members of the Mustelidae family. Along with the least weasel (below, top), members of this family include the short-tailed weasel (center), and the marten (bottom). Left: a weasel skull with its thirty-four teeth.

▼ In April or May, weasels give birth, having four to eight young. Sometimes a second litter is born during the summer. The babies weigh .5 ounces at birth and reach their adult weight four or five months later. They open their eyes at four weeks of age.

The Little Owl

A day owl!

The sun is shining brightly in a cloudless blue sky. In such ideal weather and light you would hardly expect to find an owl around! And yet, there in the meadow, on a lone apple tree, sits a little brown owl with big yellow eyes. It surveys its surroundings attentively and suddenly swoops down to snatch up a cricket that was jumping in the grass. Little owls are not at all wary of daylight.

More than 2,000 years ago, the Greeks made the little owl a symbol for Athene, their warrior goddess of great wisdom. Perhaps this was because they noticed what appears to be a thoughful look on the owl's face. The scientific name for the bird, *Athena noctua*, reflects this Greek heritage.

The little owl makes high-pitched cries that are rather happy sounding. Its cries can be heard at nightfall and also, primarily in autumn, in mid-afternoon. This owl flies much like the woodpecker—in a slow wavelike motion, leaving one perch to disappear into the hollow of a huge willow tree. Now that many of the old fruit trees have been cut down, and hedges pulled up, fewer of these owls can be seen. Unlike other kinds of owls, this owl does not nest in the woods. The little owl does not live in the United States. It can be found throughout Europe, in the Mediterranean area, and on the western side of the Red Sea.

Rather fearless creatures, little owls perch on rooftops or chimneys before taking off to hunt. Sometimes they will set up house in an abandoned building, but will not go so far as to lay eggs in a hayloft like the barn owl. Half of the little owl's diet is made up of insects, especially beetles. It will also eat small mammals such as rodents, some invertebrates, and even birds.

Length: 8½ to 9 inches.
Weight: 5 ounces.
Wingspan: 23 inches.
Lifespan in the wild: about 9 years.
Lifespan in captivity: 18 years.

▲ On the lookout from a tree, the little owl has spotted a field mouse moving through the grass. It takes off, and silently swoops down on its prey. After catching it, the little owl takes its prey back to the tree and eats it whole. Later, after digesting the food, the owl will throw up a little ball of undigested bones and hair. By examining the contents of these "balls" researchers have been able to learn more about the bird's diet.

▼ Like other nocturnal (night time) predators, the little owl's beak is hidden by feathers. Its large yellow eyes give it good vision at twilight. The large tailfeather (below) is brown with white wedges and is about three and half inches long.

▼ In summer a major portion of the little owl's diet is made up of earthworms. This little owl is pulling a worm out of the ground. Out of two hundred beakfuls of food fed to young little owls, more than half have been seen to be worms.

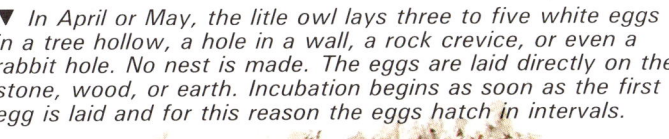

▼ In April or May, the litle owl lays three to five white eggs in a tree hollow, a hole in a wall, a rock crevice, or even a rabbit hole. No nest is made. The eggs are laid directly on the stone, wood, or earth. Incubation begins as soon as the first egg is laid and for this reason the eggs hatch in intervals.

The Ladybug

Good protection

Warblers offered ladybugs 760 times turned them down. A robin took one and then dropped it. Another robin continued to rub its beak after it had squashed a ladybug. The bird refused to have anything to do with another for eight months afterward because the experience was so bad. A song thrush was given two hundred ladybugs over a two month period and only ate eleven of them. Why were all of these birds so indifferent to such appetizing-looking beetles? The ladybug's bright colors warn predators of the sour, strong taste of its blood which comes out of its joints when it is disturbed. For this reason the ladybug is quite well-protected from most predators. However, these characteristics do not always protect these bugs. There are some small birds, like the gray flyeater, that will down ladybugs quite enjoyably.

Yellow, orange, and red ladybugs and their grubs (the young insects) feed on plant lice in fields and gardens. Some plant lice defend themselves by secreting waxy threads that get stuck in the jaws of the grubs and take a long time to clean off. However, when the ladybugs manage to attack the lice from the sides or the front, this does not happen. Above, four ladybugs are feeding on lice that are sucking plant sap. The two-spotted ladybug can eat thirty to forty lice a day. Its grubs can eat twelve to fifty.

Two spots or more

Ladybugs have a bulging shape. Their two hard wing cases are the most highly colored parts of their bodies. Under the wing cases are two membranous wings that enable them to fly. Ladybug species can be identified by the number of spots on the wings. It was once thought that the number of spots on the wings indicated the age of the insect.

Length: up to $2/5$ of an inch long, depending on the the species.

▲ On the left, a ladybug has taken flight—lifting its underwings, outer red forewings, and opening the rear wings, the bug rises up into the air. Pushed by the wind, the ladybug sails through a flock of barn swallows (center) and is caught by one of the birds. Swallows hunt in flight and eat all the insects they find. They can't tell which insects are harmful to plants and which are not. Ladybugs are not.

▼ In winter, ladybugs hibernate in compact groups. In late autumn, they break up into colonies of as many as one hundred and prepare to spend the winter in shelter provided by a plant, pole, or in the foliage on the edge of the woods.

▼ Ladybugs are red or yellow with black spots. Below are four different species. They have short, strong jaws and little antennae. Bottom row: their legs have five segments but only three are visible here. The others are hidden under their bodies.

▼ The metamorphosis of the ladybug. The female lays one to two hundred eggs. A: larvae in groups of nine to fifteen. B: the larvae hatch about ten days after being laid. C: a fully developed larva. D: the larva turns into a nymph and will be fully grown in a few days.

The Colorado Beetle

A Destructive Pest

The Colorado beetle, also known as the potato bug, was first found in America in 1823. Originally from Mexico, the beetle moved north into the Rocky Mountain area of Colorado and the western United States. There, a plant in the potato family, buffalo bur, grew in great abundance. The Colorado beetle fed on this until settlers to the area began planting potato crops. Then it began to feed on these crops. By 1860 it had spread eastward as far as Nebraska. Fifteen years later these potato bugs could be found right up to the coast of the Atlantic Ocean.

The Colorado beetle has settled in all areas where potatoes are grown, for potato leaves are its favorite food. The beetles lay their eggs on the underside of potato leaves in the spring. The chubby, ravenous larvae rapidly devour potato leaves and stem tips preventing the plant's tubers from developing.

Each season the potato bug can have two or even three broods which reach maturity in about five weeks. This rapid reproduction cycle makes these insects very dangerous to crops. After an invasion of these beetles a field can look like it has been destroyed by a hail storm. Farmers use sprays and dusts to protect their crops from the Colorado beetle.

A large family

The scientific name for the Colorado beetle, *Leptinotarsa decemlineata*, is an accurate name for the insect, describing its heart-shaped lower legs and black-striped yellow back. The Colorado beetle is part of the large Chrysomelidae (leaf) family of beetles along with the ground beetle, the Junebug, the ladybug, and the stag beetle. There are 34,000 species of beetles in the world. The largest are just over an inch long and can be found in tropical regions.

Length: 3/8 of an inch.
Lifespan: about 2 years.

▲ On the left, larvae and adult Colorado beetles on the branch of a potato plant. The larvae hatch from eggs one or two weeks after being laid. In four weeks they burrow into the ground (center) where they become nymphs. A week or two later a fully-developed adult beetle emerges and can then eat up (right) all the leaves of a potato plant.

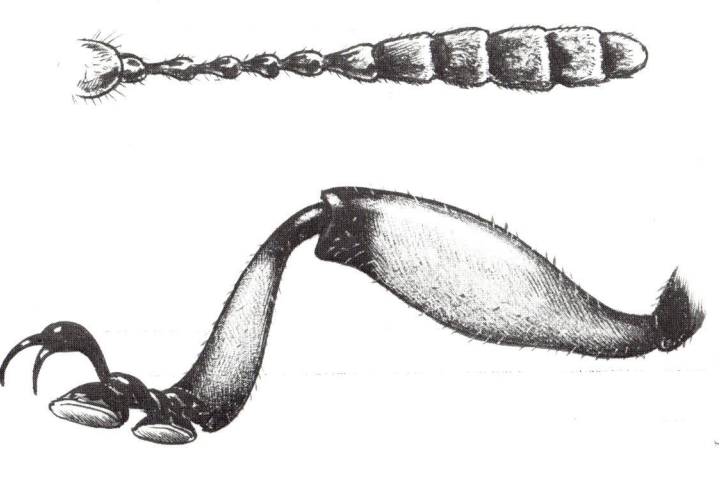 

▲ Top: Colorado beetle antennae. There are two antennae on the head of the insect. They contain several highly sensitive organs including the olfactory (smell) sense. Bottom: the leg with claws and the last segment of the foot with two cup-like bumps.

▲ This female Colorado beetle is laying her eggs. Eggs are laid in groups of twelve to eighty on the inside of potato leaves. In June, the insect lays from seven to eight hundred eggs which will hatch in the autumn. The adult beetles spend all winter underground.

▼ The gray partridge is one of the enemies of the Colorado beetle. The gray partridge, however, only eats the insects during the three months of spring. In fact, baby partridges are the ones who eat a great many small animals and Colorado beetles are part of their diet.

The Starling

Large flocks

Estimates have shown that the weight of all the starlings found over five acres of land would be as much as eighty tons! Starlings have always been common in Europe but were only introduced into this country in the late nineteenth century. Then, one hundred starlings were released in New York City. Today there are millions of starlings in North America, descended from these first one hundred. They can be found throughout most of the United States and in southern Canada. Starlings gather in very large flocks. Researchers in France identified 281 separate flocks of starlings. Thirty of those flocks contained at least one million birds!

During the day the birds split up into groups of thousands and fly over fields and meadows to look for food. They are not afraid to go near farms to do so. A starling needs about one ounce of food per day, this is about half its weight.

In spring starlings feed on many insects that are harmful to trees, crops, and meadowland. Although starlings help farmers, they are also an annoyance because they eat up fruit and grain crops. They are particularly fond of cherries.

However, fruit, which is only available for a short time, accounts for only a small portion of the starlings' diet. Depending on the season they also eat lots of insects and seeds.

Continental travelers

In Europe, when the young leave the nest, they fly around in all directions before gathering in flocks to migrate for the winter. Every year millions of starlings from Germany, Switzerland, Czechoslovakia, Poland, and the Soviet Union fly to France, where the climate is warmer, for several months. They leave again in early spring. In North America, however, most starlings do not migrate.

Length: 7½ to 8½ inches.
Weight: 2 to 2.8 ounces.
Wingspan: 15½ inches.

▲ Above left: these starlings perched on power lines are about to take off. Top center: their flock, at first it looks like a great ball and then it unwinds like a ribbon. How do so many of these birds manage to fly together? What secret arrangement do they have to keep from colliding with each other in mid-air? It is believed that certain birds in the flock are responsible for changing flight paths, and the others follow them.

▲ A starling's beak, brownish gray in autumn and yellow in spring, is about an inch long. The bird's feet are typical of members of the Passeriformes order, that is, birds that perch. It has three claws in the front and one in the rear. Under the tail the large feathers cover smaller, softer underfeathers.

▼ Coming from twelve to eighteen miles away, hundreds of thousands of starlings return "home" to the same "dormitory" every night. At nightfall, sometimes more than a million of these birds will crowd into a group of trees. In the morning

▲ The female usually lays four to six pale blue eggs in a nest made of dried grass and roots that she has built in a tree hollow or hole in a wall. Incubation takes twelve days and fledglings first leave the nest when they are about twenty-one days old. Above: a mother feeding her young.

they split up into smaller flocks and fly off, leaving a thick layer of dung on the ground below. Vegetation on the ground disappears due to the burning effect of their droppings.

The June Bug

Two years underground

What a strange life cycle June bugs (also called May beetles) have! For two years or more June bug larvae live underground feeding on all kinds of plant roots. In August or September, they develop into winged adults but stay hidden underground. In spring they finally come out of the ground. After such long preparation, they live for only a few more weeks. The adult June bug only feeds on tree or flower leaves for about fifteen days before reproducing itself and then dying. So it is the larvae, called white grubs, that have the longer lifespan of the two. June bugs only appear in great numbers every three years.

Rootless crops

The common June bug attacks the roots of grain, alfalfa, and vegetable plants. The larvae cause the most damage to crops in their second year underground. The amount of damage caused by June bugs depends on the density of plants to larvae. There have been cases when entire rows of vegetables have been destroyed because their roots have been eaten off. At night adults feed on the leaves of the following trees: oak, elm, beech, maple, poplar, willow, and walnut. They are also fond of fruit tree leaves. We usually see these bugs at night when light attracts them.

There are several different kinds of June bugs in the United States—the hairy May beetle, the fuscous May beetle, and the prairie May beetle. Most are reddish brown in color, except for the hairy May beetle which is a brownish yellow.

Length: .8 to 1.2 inches.
Larva's lifespan: 3 to 5 years.
Adult's lifespan: about 11 months

▲ In May, the June bug lays eggs four to five inches underground in meadows, alfalfa fields, and even under saplings. Above left, the female is shown laying what will be sixty to eighty eggs which hatch in four to six weeks. Above center, a larva, called a white grub is gnawing on a root. In summer, June bug larvae burrow down as far as two feet to avoid dry earth. Top right: the larva has changed into a nymph.

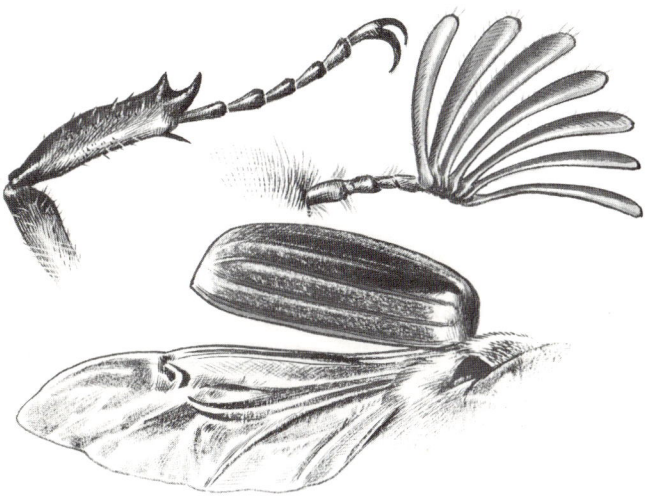

▲ Above left: one of the June bug's six legs. The five little segments (the last of which has a claw) are called the tarsus. Right: an open antenna. A male's antenna has seven "flaps" and the female's, six. Below: the wing cases protect the membranous underwing.

▲ When there are too many larvae in one place, for example, seventy to a square yard, food becomes scarce and many die. Here, a fully grown larva two inches long is devouring a smaller one since no other food is available.

▼ Like the ladybug, the June bug has two hard wing cases and two fragile, thin, rearwings. To fly, it lifts the forewings and moves the rearwings. The June bug flies mostly at night to feed on trees. Its sense of direction leads it to the edge of woods.

The Brown Hare

When chased by hunting dogs, a hare can run for several miles at a speed of more than thirty miles per hour. For shorter distances it can easily move at fifty miles an hour. When threatened, the hare takes off like a shot and can confuse its trackers by zigzagging and retracing its steps to confuse its pursuer's sense of smell. A hare can reach these high speeds because it has very long hind legs and a large heart. On downhill runs, it can jump surprisingly far. On snow a hare has been seen to make jumps almost ten yards long. This quick speed combined with the hare's long, mobile ears make it well-adapted for sensing and escaping danger.

The brown hare pictured below is common to the plains of Europe, the steppes of Africa, and parts of Asia. It has been introduced into the United States from Europe and can be found in the Northeastern states. Its size and brownish-gray color vary slightly depending on the continent on which it lives. The brown hare can be found on cultivated land, fields, and dry ground. Vegetation on the ground must be short for the hare to be able to run rapidly over it.

The hare goes into the woods mostly in winter to look for some of the food it eats, especially plant buds. It gnaws on the shoots of young trees and chews the bark off trunks, causing serious damage to trees.

Unlike rabbits, who live together in underground warrens, most hares lead solitary lives, living on their own above ground, for a good part of the year. There are some hares that have been known to burrow, but only as protection against extreme weather conditions—the intense summer heat of the desert, or the harsh cold of winter snows.

Foxes and weasels are a leveret's (young hare's) main natural enemies. Adults are killed by hunters or tularemia (rabbit disease). A hare usually spends its entire life on the same five to twelve acres of land.

Length: head and body—25 to 27 inches; ears—4½ to 5 inches.
Weight: 7 to 10 pounds.
Lifespan in the wild: up to 12 years.

▲ From late winter until July or August, hares come together for mating and the males box with each other. In spring, five, six, or even more hares can frequently be seen together in fields. Top left: a male brown hare and two females. Center: a male pursuing a female. Top right: two males fighting over a female, pulling out tufts of each other's fur, and scratching one another.

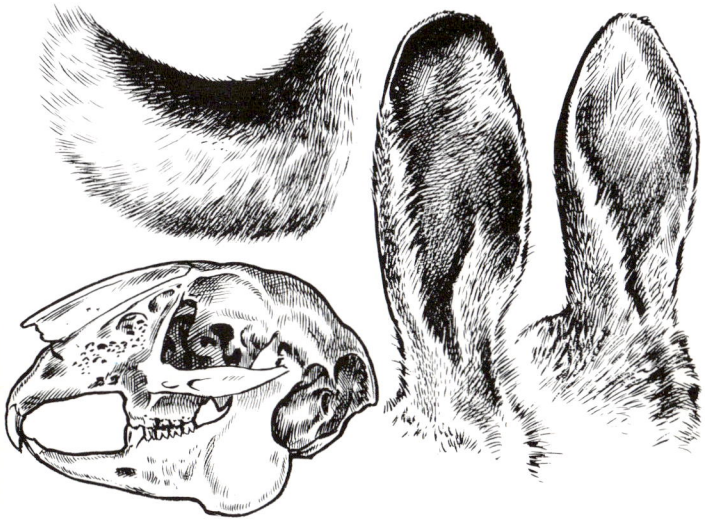

▲ The hare is larger than the rabbit. Its ears are longer (left: a hare ear; right: a rabbit ear) and have black spots. The hare's back legs are more developed. The four incisors in the upper jaw make the hare different from a rodent which has only two. Top left: a hare's tail.

▲ Unlike baby rabbits (kittens), leverets are born with a coat and opened eyes. They can walk and hop just after birth, but usually stay in a sheltered place while their mother suckles them. At birth they weigh about three ounces. A female can have three to four litters a year of two or four leverets each.

▼ During the day, hares dig a shallow hole in the ground to fit their body into. In this improvised shelter (hares never make underground warrens or tunnels) hares keep their hind legs folded under them. In this way they can always be on the lookout and ready to spring away if they feel threatened.

The Stonechat

An observer

Perched on a fence post the stonechat, a kind of thrush, surveys the path looking for a caterpillar or grasshopper. The moment it spots one, it goes after it immediately. The stonechat has been observed to have spotted a half-to three-quarter-inch prey ninety feet away and snatch it up. The bird in the picture on the left has captured several insects, probably to feed to its young. It also eats mollusks and, in the winter, seeds.

The stonechat spends a good deal of its time perched, watching the world around it. This bird can be seen on the side of the road, on telephone wires, in bushes, or in fields. It is easily recognizable by its black head, red breast, and the white collar-like marks around its neck. When approached, the stonechat shows its fear by beating its wings and making harsh cries. It is not, however, afraid of cars or trains and frequently even nests on railroad track embankments. This bird is found throughout Europe, but not in the United States.

A strange song

The stonechat's name comes from the sound of its voice. The bird's song resembles the sound of pebbles being rattled together.

In autumn the stonechat moves from the colder, northern areas of Europe south to western France, the Midi region, and sometimes as far away as Spain. In Brittany (France), it is a sedentary bird—it does not migrate. Some stonechats try to survive winters in eastern France but perish from the cold and lack of food.

The stonechat is not found in developed areas. This is because it prefers to be near fallow or uncultivated land.

Length: 5 inches.
Weight: ½ ounce.
Wingspan: 7 inches.

▲ The mating dance of the stonechat takes place in March. The male who was perched (above left) in a bush, flies straight up for about forty feet and then flies down again several times like a mechanical toy while beating its wings rapidly (center). Above right: the male pursues the female. He will then land near her and spread his tail feathers, rhythmically moving them up and down while his wings are spread.

▲ Above top left: the male stonechat. Above top right: the male wingchat. The wingchat is a migrating bird that spends its winters as far south as tropical Africa. Bottom left: the stonechat's beak is about half an inch long. Bottom right: the white band on the wing has small feathers that cover the base of the longer feathers.

▲ The female builds its nest out of grass and moss. She usually builds it in a sheltered hollow in the ground and then lays five or six eggs in it. The eggs are about two-thirds of an inch long and three-fifths of an inch wide. Hatching occurs about fifteen days later.

▼ This male stonechat caught a peacock butterfly and will take it to the fledglings in the nest. After the young leave the nest for the first time, at fifteen days, they continue to be fed for another twenty days or so by their parents.

Spiders

The garden spider is a medium-sized, web-spinning spider. It can be found in the northeastern United States and in Europe. The garden spider's web is a geometric work of art and a fatal trap for many insects. The spider's web is made of silk threads. These come out of tiny holes on the spider's body called spinnerets. The threads are very fine, a fraction of a millimeter in diameter, yet strong.

The spider spins sticky threads on a framework of spokes. These sticky threads trap insects. For the trap to work properly the threads have to be kept clean of dust and respun if broken. The spider eats the old threads, for they are a good source of protein.

When an insect gets caught in the web a spider can determine its size and location by shaking the threads of the web. Fine hairs at the tips of the spider's legs are sensitive to vibrations and tell it of the presence of an insect in its web.

In the picture at left a wasp has become caught in a web. First, the spider injects poison into the wasp. This keeps the wasp still and breaks down its insides. Next, the spider wraps the wasp in silk threads and carries it to the center of the web. Now, the spider eats by sucking the juices out of the wasp.

Insects have six legs and bodies made up of three sections. Spiders have eight legs and bodies with only two sections. Spiders are not insects, but arachnids. On the end of each of the spider's legs are two comb-shaped claws. These are used for cleaning the web and help the spider move around. Between its eyes and mouth the spider has two pincer-like fangs. A little canal at the tip of each fang leads to the poison gland. A spider can have as many as eight eyes. Most web-spinning spiders have poor eyesight, though, and rely only on their webs for catching prey.

The garden spider is just one kind of spider. Throughout the world there are over 30,000 species of known spiders. They can be found everywhere in the world except in Antarctica.

Size: up to one inch.

▲ The garden spider begins construction of its web by spinning a thread between two supports. Usually it relies on the wind to carry the first thread from one support to the other. From this thread it spins a vertical thread. These first threads form a Y-shape. Then the spider spins the spokes of the web out from the center. Last, it spins sticky threads in a circle in which its prey get stuck. All this takes about an hour. A garden spider's web can be three to four feet wide.

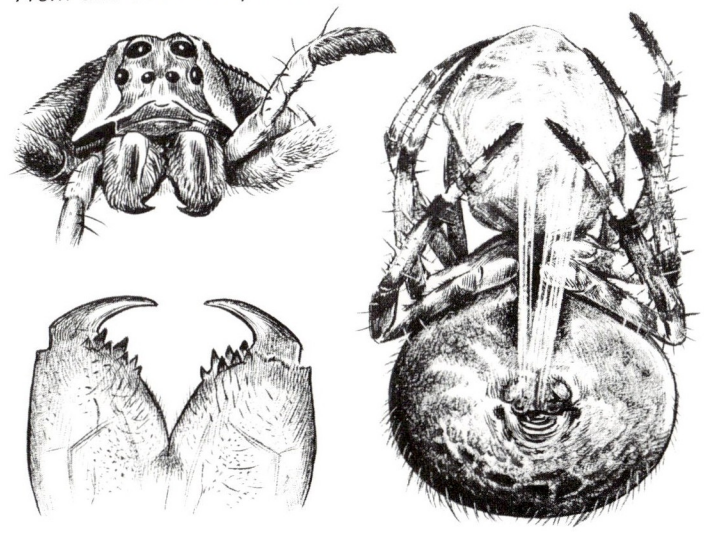

▲ Above left: an enlarged view of the head of a hunting spider, with six of its eyes. Below left: a close-up of the fangs the spider uses for defense, seizing prey, cutting thread, and crushing food. Right: a garden spider viewed from underneath while it wraps its prey in silk threads.

▲ This spider is carrying its young on its back. The eggs are covered by a soft cocoon. Not all spiders do this, though. Some spiders attach the cocoon to plants and then abandon it. Others stay near the cocoon or carry the eggs around until they hatch. Garden spiders hatch in May.

▼ Male spiders are usually smaller than females. Here, a male approaches a female to mate. He must be cautious for frequently a female spider will mistake a male for an enemy and, instead of accepting him as a mate, kill him and eat him!

Snails and Slugs

Leaning to the right and then to the left side of its "foot," the snail can cover a maximum of three inches a minute. The snail slides easily over rough terrain because the glands toward the front of its body emit a thin film of mucus. This reduces friction and allows it to move along easily. Snails are mollusks, and part of the group called gastropods. This name means "belly-footed" and refers to the wide foot on the snail's underside, or belly.

The snails pictured above have come together to mate. Before doing so, they will spend hours crawling over one another. Land snails are hermaphrodites—they have male and female sex organs. Each snail will stick a dart into the other to make their reproductive cells function.

The snail can withdraw into its shell if it feels threatened by other creatures. Its shell also provides shelter in dry or cold weather. In those times it seals the shell opening with a layer of calcium-filled mucus which dries and hardens, but still allows air to get through. This "door," called an epiphragm, keeps the snail's body (which is 84% water) from becoming dehydrated (dried out). The snail can make up for any loss of liquid by absorbing dew or rainwater through its skin. Land snails generally live in damp places.

The snail uses it's radula, a ribbon-like organ with rows of horny "teeth," to scrape up food. Some kinds of snails have thousands of these tiny teeth. With its radula the snail can scrape up bits of leaves, fruits, and mushrooms. Snails are called scavengers because they eat dead animal and plant matter. The snail's olfactory (smell) organ, used to find food, is in the horns of the radula and can detect food eight inches away.

Shell diameter: 1 to 2½ inches.
Weight: half an ounce or more.
Lifespan: 6 to 7 years.
Length of the slug: 1 to 2 inches.

▲ Slugs are another kind of gastropod. They are snails with no shells. A slug has no real appendages but can get around what may seem to be insurmountable obstacles by using the rather thick, sticky mucus produced by glands at the rear of its body. Above: a slug moves from a twig to a stone below. It puts out a thread of mucus, hangs on it as if from an elastic band, and slides down to the stone.

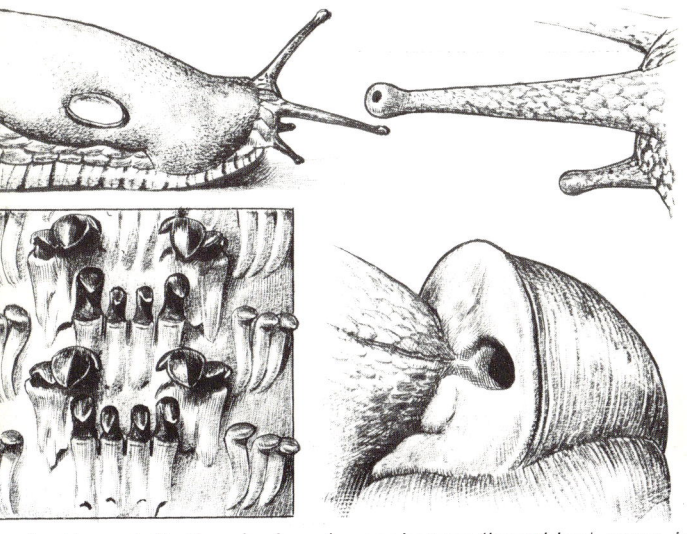

▲ Above left: the slug's only respiratory (breathing) organ is on the side of its body. Above right: the slug's eyes are at the tips of its longer tentacles. Below left: part of a snail's radula enlarged many times. Below right: the snail's respiratory organ.

▲ Six to eight weeks after mating the garden snail lays twenty to sixty eggs in a hole in the ground. It closes up the opening and abandons the eggs. When hatched, the baby snails already have tiny shells. They grow slowly and can only reproduce at three years of age.

▼ Some ground beetles like to feed on snails. This one has long legs that help it move quickly. It also has powerful jaws. Faced with this formidable enemy, the snail has withdrawn into its shell to protect itself. The snail secretes a foam-like substance to keep the enemy from moving any closer. Ground beetles attack mainly small snails.

The Earthworm

There are, on the average, from five to twelve million earthworms in two and a half acres of meadowland. In a cultivated field, however, it is estimated that there are hardly more than four million on the same amount of land. A clue to their density is provided by the number of worm holes that can be found on the surface of the ground. As many as 1,000 have been counted to the square yard, with each worm boring more than one hole. The presence of worms can also be detected by the little twists of excrement they leave behind. In this way they manage to shift a considerable amount of earth. The weight of these twists varies from four to sixteen pounds per square yard or twenty to eighty tons over two and a half acres.

These somewhat unattractive creatures play an important role in nature. By digging their underground tunnels they aerate (create air pockets in) the soil. This makes it easier for the soil to absorb rain water and leaves air spaces for the roots to grow in. They also help to make humus by plowing vegetable matter, which they get at the surface in damp weather, back into the soil. They eat decayed vegetable matter, bits of mushrooms, and humus.

If an earthworm is cut in half, the halves do not become two separate worms. An earthworm can grow a new tail or head only if the cut has been made near the end of its body. So, the earthworm's powers of "reproduction" are not as extraordinary as is sometimes thought. The earthworm's existence is full of all kinds of hazards. They are frequently mutilated by spades and plows as the earth is turned over. They die in autumn freezes or summer heat if they are not deep enough into the ground. There are some kinds of beetles that eat up the young worms just after they have hatched from their cocoons. Earthworms are a delicacy for small animals, like moles; and a staple food source for many birds, who are their natural enemies.

Length: 4 to 8 inches.
Weight: ½ an ounce or more.

▲ The earthworm moves by contracting its vertical and horizontal muscles. Top left: the bulging center of the body grips the ground with tiny bristles while the front part moves forward. Top right: the front part takes over, pulling the rear part of the worm along. Bottom left and right: the rear contractions push the front part of the worm forward. According to some estimations, the common earthworm can cover sixty-five yards in an hour.

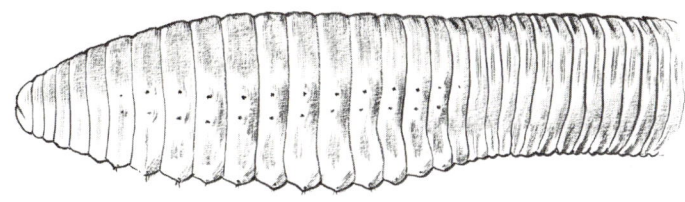

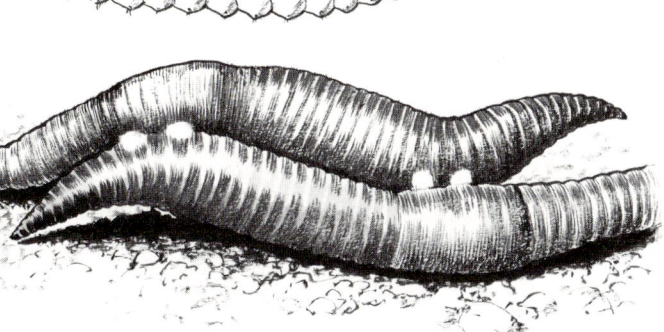

▲ Top: the front part of the worm with tiny bristles. When viewed through a microscope, each species can be identified by its unique bristles. Bottom: mating worms are held together by strands of mucus. Each worm has male and female organs but they must join together for fertilization to take place.

▼ This worm will not escape from the toad and the hedgehog. Tawny owls hunting at night in meadowland also prey on worms. While observing the tawny owl through infrared binoculars it was found that the bird can eat up to seven earthworms in an hour.

▲ In manure and compost heaps (slowly decaying vegetable matter), worm density can be very high. These are not, however, the same species that are found in meadows. The most common species of worm varies in color from pink to brown and is two to five inches long.

The Toad

A gardener's friend

The toad likes to use the same hiding place for months or even years. It deserves to be called the gardener's friend because during the summer it gobbles up all the little creatures it can find at ground level. Very often these creatures—caterpillars, slugs, and small snails—are harmful to lettuce and other plants. Toads are not selective, they eat creatures that humans find useful or harmful. The fact that they consume slugs makes them generally well liked, even though their appearance could be called unsightly.

The toad's favorite time for hunting is at nightfall and during the night. Toads like to eat worms, insects, and baby frogs but they don't like long caterpillars with stiff hairs. As many as 363 ants have been found in the stomach of a toad.

In harsher climates toads are only seen in spring and summer. In the winter they find a hole made by another animal and take shelter there from the cold. They begin to hibernate in October or November.

Piggyback toads

During the mating season the male toad puffs out the balloon-like part of his throat to make his mating call. When he has attracted a female he climbs onto her back and she carries him to the pond where she will lay her eggs. He holds on by digging his forelegs under her forelegs. If another male tries to push him off he will kick him with his hind legs. However, other males do sometimes manage to hang onto the female as well. A female is much larger than a male and may sometimes carry as many as ten suitors on her back! When this happens there is a chance she could be crushed to death by their combined weight.

Length: 2 to 5 inches.
Weight: 1 to 2½ ounces.
Lifespan: 10 years in captivity. 1 to 2 years in the wild.

▲ This toad has spotted a fly (left). Will he pounce on it? No, he doesn't catch his food that way. The toad has a long, sticky tongue that can be a dangerous weapon. Its tongue is about three quarters of an inch long, thick, and rounded. He flicks it out and scoops up his prey and swallows it in exactly one tenth of a second. The fly is caught and swallowed before it is able to react. Toads only hunt small live creatures.

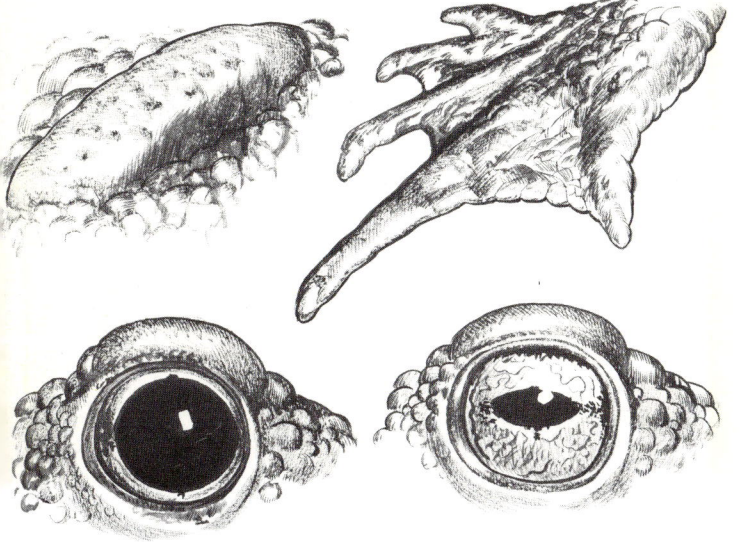

▲ Top left: the toad's largest warts, above its eyes, are actually glands containing a poison that the toad releases when attacked. Top right: the end of a leg. The forelegs have four fingers and the backlegs have five. Bottom left: at night, a dilated pupil. During the day it is contracted (bottom right).

▼ Tadpoles and young toads are food for many creatures, including herons, carrion crows, polecats, skunks, and otters. In this country the hognose snake is a mortal enemy of an adult toad. This snake is one of the few creatures that is not

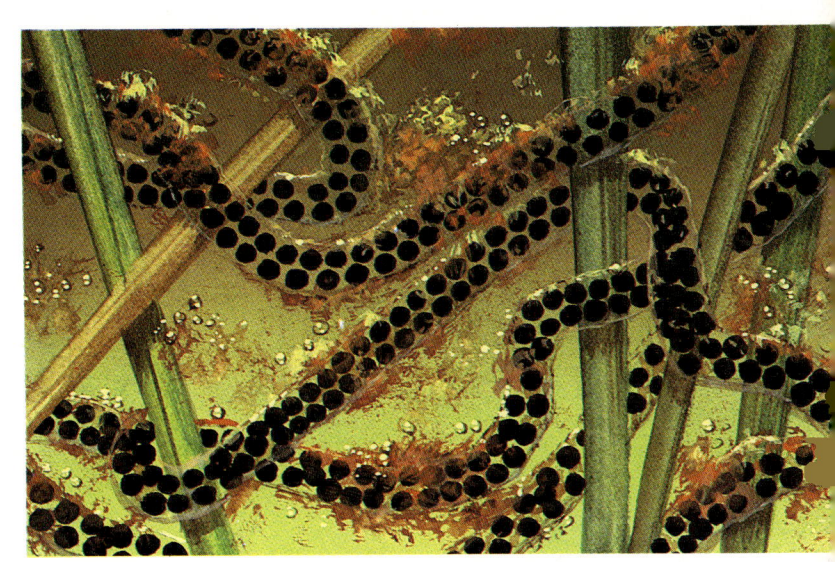

▲ In spring, the female lays 2,000 to 3,000 eggs in a pond. The male clings to her and fertilizes the eggs. A clear jelly covering protects them and keeps them together. The eggs are in long strings that cling to aquatic plants. In ten days tadpoles emerge. They grow into toads in two or three months.

affected by the toad's poison. In the presence of an enemy, the toad can look quite fearsome by rising up on its hind legs and puffing up its body.

The Blind Worm

An uncanny resemblance

Legless, the blindworm is often mistaken for a snake. It is absolutely harmless, rarely bites, and its teeth are not connected to poison glands. Its habit of sticking out its tongue all the time also makes it seem like a snake. In addition to humans, it has many other natural enemies—hedgehogs, grass snakes, badgers, foxes, rats, and white storks. Many birds of prey—the kestrel falcon and the red kite—hunt blindworms regularly. In spite of these dangers the blindworms manage to remain quite abundant by spending much time sheltered by vegetation, and hiding under rocks and in woodpiles. They can also dig holes for themselves. At daybreak they come out of hiding to look for food. In spring, at mating time, they can more frequently be seen out in the open during the day.

A misleading name

The blindworm slithers along the ground like a snake, but more slowly. Strange as it may seem, this creature is not a snake or worm, but is actually a legless lizard. Like other lizards it has a movable lower eyelid and eyes, so it is not blind by any means.

Blindworms feed on spiders, insects, earthworms, snails, and slugs. They have been seen to gobble up seventeen small slugs at a time! They only like moving prey and ignore dead or immobile animals. Blindworms live in fields, woods, hedges, and gardens throughout most of Europe. They are not found in the United States but a similar legless lizard, the glass snake, does live here. In October, blindworms slither into leaf covered burrows to hibernate, either alone or in groups, until spring. Twenty to thirty have been found hibernating together.

Maximum length: 1 foot. The females are longer than the males.
Lifespan: unknown in the wild, up to 50 years in captivity.

▲ The blindworm molts (sheds its skin) periodically like snakes and lizards. Above left: the old skin is being shed from the snout down. The head swells to help the process get started. The blindworm rubs its body on the ground to speed up the process (right). Sometimes the skin comes off in one piece but more frequently it is in long strips.

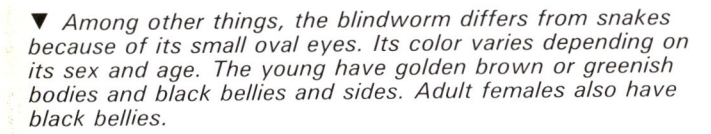

▼ Among other things, the blindworm differs from snakes because of its small oval eyes. Its color varies depending on its sex and age. The young have golden brown or greenish bodies and black bellies and sides. Adult females also have black bellies.

▼ Picked up by the tail, the blindworm escapes by breaking in two! One of its vertebra, which is already cracked, breaks at the slightest pressure. When it feels pressure, the blindworm contracts its muscles and the end of the tail breaks away from its body.

▼ Blindworms are oviparous. That means their young hatch from eggs outside the mother's body. The young are born in August or September in groups of six to twelve. They grow slowly from two to three inches at birth to nine inches at two years.

The Countryside Through the Seasons

The landscape of the countryside, as we know it today, is the result of patient work carried out by many generations of farmers over the course of almost two thousand years. At one time, much of the landscape was covered with thick forests. Little by little, trees were cut down and the land was cleared to make fields and meadows. People began to replace part of the natural vegetation with plants which they used to feed themselves and their animals or to make textiles.

A Variety Of Sources

In meadows, people usually made use of indigenous plants but learned how to mix various species to obtain greater nutritional value for animal feed. Among these species are alfalfa and clover. Other species that were planted in fields came from many different places. In Europe, corn, potatoes, and tomatoes were brought from the Americas. Lettuce and cabbage already existed there but were improved through selection and crossbreeding.

Other species that were not useful to people also grew in meadowland. These "weeds" had to be destroyed or pulled up since they threatened the existence of the useful plants. Some weeds are indigenous to a certain area while there are often others that have been introduced accidentally in trading with countries overseas.

Changing landscape

The environment of the meadowland remains relatively stable since weeds and other plants growing in them are rarely pulled up. In summer, these meadows are mown several times. Fields, however, are cleared and plowed after every harvest of the crops that grow in them such as wheat, corn, or potatoes. The land therefore looks different depending on the season and the year.

In winter, many fields are bare except for ones where rye, winter wheat, and barley grow. In summer, corn, beets and other varieties of barley and corn are sown, and potatoes are planted. Although many types of grain appear to be the same, they are different in color and size. Rye, for example, grows taller than wheat and some types of wheat have "whiskers". Flax has blue, rectangular flowers in June and colza is bright yellow. In summer, grain has the color of dried hay and after harvest time, only its roots and stubble are left before plowing.

After beets and corn are harvested the vegetation practically disappears from the fields which are then prepared for the next planting. All year long, farmers work on their land, either sowing or harvesting. In spite of the progress machines have brought to farming, farmers are still dependent on the weather. In dry weather, a farmer can water the fields but is rather helpless against freezes and an over abundance of rain. Certain crops flourish certain years depending on the weather. Beets and potatoes grow best in damp weather. Farmers can never be sure of the size of their harvests since they have to deal with the rather unpredictable forces of nature.